Weapon of Choice

The US Navy's Role in Forging Trinidad and Tobago's National Cultural Identity 1940-1947

Louis Miranda Diaz

Copyright © 2022

Comments & Recognitions

U.S. Southern Command
9301 NW 33rd Street
Doral, FL 33172-1202

June 25, 2025

Dear Mr. Miranda,

Thank you for the generous gift of the book *Weapon of Choice – The US Navy's Role in Forging Trinidad and Tobago's National Culture Identity.* I am grateful for your support of our military community. Knowing that thoughtful individuals like you are keeping us in mind makes all the difference.

Again, please accept my deepest thanks for your generosity and continued support.

Sincerely,

ALVIN HOLSEY
Admiral, U.S. Navy
Commander, U.S. Southern Command

OFFICE OF THE PRESIDENT
REPUBLIC OF TRINIDAD AND TOBAGO
Circular Road, St. Ann's, Port of Spain
Telephone: 1(868)225-4687, 625-9815
Fax: 1(868)627-9886 E-mail: otp.mail@otp.gov.tt Website: www.otp.tt

1 November 2023

Mr. Louis Miranda
Via email: lmiranda1121@gmail.com

Dear Mr. Miranda,

I have been directed by Her Excellency Christine C. Kangaloo O.R.T.T., President of the Republic of Trinidad and Tobago, to respond to your correspondence by email dated 27 October 2023.

Her Excellency thanks you for your letter and your offer of a gift of your book, *Weapon of Choice: The U.S. Navy's Role in Forging Trinidad and Tobago's National Cultural Identity.* Her Excellency congratulates you on its publication and she would be happy to place the text in the library of The President's House so that it be available for the enjoyment of future Heads of State and their families.

Her Excellency sends you her best wishes. She looks forward to reading your book.

Sincerely,

Kent Jardine

Kent Jardine
Executive Adviser
Office of the President

The Library of Congress
U.S./Anglo Division
101 Independence AVE SE
Washington, D.C. 20540-4270

June 27, 2023

Louis Miranda
Lmiranda1121@gmail.com

Dear Mr. Miranda:

On behalf of the Library of Congress, we want to thank you and acknowledge the generous donation of your publication, **Weapon of choice: the US Navy's role in forging Trinidad and Tobago's national cultural identity, 1940-1946.** We are happy to inform you that the publication was selected for addition to the Library's General Collections and was assigned Library of Congress control number: **2022482027**.

Once again, thank you for your interest in enriching the collections of the Library and your most kind donation.

Sincerely yours,

Jennifer Baum

Jennifer Baum
Head, U.S. Special Acquisitions Section
gifts@loc.gov

Head Office:
The National Library of Trinidad and Tobago
Hart and Abercromby Streets
Port of Spain
Republic of Trinidad and Tobago
P.O. Box 547, Port of Spain
Republic of Trinidad and Tobago

Phone: (868) 62-NALIS (62-62547);
623-9673; 623-6962; 623-7278;
624-1130; 624-5075; 624-4466;
625-5233; 627-3679; 627-8307;
627-1878; 624-6541; 624-3409
Fax: (868) 625-6096
E-mail: nalis@nalis.gov.tt
Website: www.nalis.gov.tt

HLD Ref.: 7.2.1

August 5, 2023

Mr. Louis Miranda
Author

Dear Mr. Miranda,

Re: Donation of Book to NALIS

We wish to extend our sincere thanks to you, for your kind gift of the book '**Weapon of Choice - The US Navy's Role in Forging Trinidad and Tobago's National Cultural Identity 1940 - 1947**'.

This gift is indeed a most welcome addition to our collection. Our compliments to you for your public-spiritedness in choosing the National Library of Trinidad and Tobago as a repository for this treasured publication.

With many thanks.

Yours faithfully,

Jasmin A. Simmons
Director, Heritage Library Division
National Library and Information System Authority

Winner of the Prime Minister's Innovating for Service Excellence Award 2004

"Her Excellency congratulates you on its publication and she would be happy to place the text in the library of The President's House so that it be available for the enjoyment of future Heads of State and their families."

Kent Jardine, Executive Advisor, Office of the President

"This gift is indeed a most welcome addition to our collection. Our compliments to you for your public-spiritedness in choosing the National Library of Trinidad and Tobago as a repository for this treasured publication."

Jasmin A. Simmons, Director, Heritage Library Division, National Library and Information System Authority

"Hi, Louis, . . . thank you so much for giving me this precious gift. I'll forever cherish this book and the inspiring education I've gained reading it. Rest assured it will be shared with my kids and generations to come. It was well worth the read, very insightful thank you."

Alexis Darren, Hotel Director, Royal Caribbean Cruise Line

"Your book is fantastic, read it late but after I started was hooked onto it. Got to know so many things, never read about all this. A great book to get so much of an insight."

Rashmi Kurpad, Bengaluru, India

"I finally finished reading your book it surely was interesting because I learnt a lot more of our history and you also took me back to my childhood days when my friends and I would walk up to Waller field and play in the abandoned cold storage on the base we did not live too far away from there thanks for the nostalgia and congratulations keep up your good work God Bless"

Meela Brown, Retired, San Juan, Trinidad

"Te felicito y la verdad es que lo celebro porque has hecho un trabajo original y bien creativo que aporta al desarrollo historiográfico del caribe."

Luis Figueroa, PhD, San Sebastian, Puerto Rico

"I have read most of your book and it has a lot of interesting information about Trinidad's role during World War II."

Rajkumar Bissessar, Petrotrin, Retired

"This book would be a good addition to any history buffs Library as well as our National Archives, National Library and University of the West Indies (UWI) Library."

Claire Woods, Carnival Institute of Trinidad and Tobago

"You honored Trinidad by the telling part of their history."

Richard Miranda, USMC, Retired

"My concept of Trinidad and Tobago's calypso and steelpan music has forever changed."
Trinidad and Tobago Coast Guard Executive Officer

Preface

Having served with the both the U.S. Coast Guard and the U.S. Navy, and my travels to Trinidad and Tobago (henceforth Trinidad) for over 30-years, is what inspired me to select this subject matter. What I found most interesting is that when I asked my Trini friends if they knew the history of the discovery of the steelpan – their national instrument – they said no, that it is not taught in their schools.

The two primary sources used in undertaking this investigation were official U.S. Naval documents submitted to Congress on 3 July 1946, consisting of two volumes, titled "Building the Navy's Bases in World War II". These volumes provide details of the U.S. Navy's arrival to the Caribbean islands, projects undertaken, detailed costs, and the issues they confronted with some of the colonial governments, their population, and anti-colonial forces. Secondary sources were used to complement and substantiate the information contained in the U.S. Naval report.

This short book stems from the history master degree thesis presented as a graduation prerequisite.

Acknowledgements

The kind attention and willingness of people and institutions to help me in putting this book together is most gratifying and appreciated.

First and foremost, my dear friend Iviahoca Ramos Martinez, who always embraces my projects as if they were her own, and gives me fresh and insightful ideas. My friend Josery Moya Ruiz, who inspired me to continue writing my book by saying that the book would be the legacy I would leave behind to my family and friends. My cousin Diana Mejías, PhD, who suggested that I study my master's degree and helped me edit it. My professor at CEAPRC, Evelyn Vélez Rodríguez, PhD, who showed a keen interest in my thesis' subject matter. My brother Pete Miranda who gave me invaluable ideas. My dear friend, Captain Richard Acquavella, USN Retired, and my niece Admiral Kristin Acquavella for their steadfast belief in my project. My friend Jennette Remak, from Long Island, NY, who took time from her schedule to assist me in obtaining pictures for this book. My dear friend Alethea Alleyne, from Trinidad, who was especially helpful in gathering information and contacts for me. Dr. Kim Johnson, Director of the Carnival Institute of Trinidad and Tobago, considered to be the foremost historian and filmmaker of the steelpan, and Ms. Claire Woods, his assistant, were kind enough to take time from their busy schedule to provide insights and share their thoughts with me.

Last, but not least, my wife, Ana María Severino Trinidad, whom I could count on for her unwavering support.

I dedicate this book
to my children and grandchildren.
Bless them.

The US Navy Makes Landing

Little did the government and people of Trinidad know, that a battle for survival was taking place some four thousand five-hundred (4,500) miles away, which would forever change their way of life. It was the battle of Dunkirk (26 May - 4 June 1940), where British, French, and Belgian forces were pinned down with their backs toward the beaches of Dunkerque (Dunkirk), France. The Nazis were fast approaching, and whom vastly outnumbered the allies in forces, as well as in weapons. It was going to be a slaughter. Meanwhile, Sir Winston Churchill, the newly seated Prime Minister of the United Kingdom, was desperately negotiating with President Franklin Delano Roosevelt for destroyers to counter Germany's onslaught.

Trinidad, then a former British colony and birthplace of the steelpan, is the southernmost island of the Caribbean archipelago, which together with Venezuela forms the Gulf of Paria. In prehistoric times, Trinidad was connected to Venezuela by a land bridge. As a result, it is the only island in the Caribbean that manifests fauna and flora indigenous to both South America and the Caribbean. Few beaches in Trinidad enjoy the crystal blue waters like that of its sister island Tobago or, for that matter, the rest of the Caribbean. This is due to the vast delta formed by Venezuela's Orinoco River and its distributaries, which, at times, reaches the northern coast of the island, giving their shores a muddy river like appearance.

Trinidad has always been a traditionally rich and complex society consisting of a diversity of ethnic and religious factions, each looking out for their own cultural and political interests. By World War II (Sept 1,1939 – Sept 2, 1945), the people of Trinidad viewed themselves as having a distinct set of values from that of the British, but it was precisely this diverse social and cultural structure that did not allow them to develop a much-needed national cultural identity. Hence, this rich diversity made any effort in uniting the population as a nation, as "Trinis", was difficult at best. The U.S. Navy's arrival during World War II played an unwitting, yet important role, in Trinidad forging this identity.[1]

The 1946 census showed that of a *"total population of five hundred fifty-seven thousand nine hundred seventy (557,970), 35.1 percent were of Indian descent, 46.9 percent were of African lineage, 14.1 percent were of mixed race, 2.7 percent were white, 1 percent were Chinese, 0.2 percent were of Syrian, Lebanese, or Arab origin, and 150 were of unknown descent."*[2] Each, of course, had their own customs, form of dressing, religious beliefs, diet, as well as musical, social and political interests. While the census of 2008, showed that *"the largest religious group is Christianity with 63.2 percent of the population. This includes Protestant Christians (with Anglicans, Presbyterians, Methodists, Evangelicals, Pentecostals, Shouter or Spiritual Baptists and regular*

---

[1] This article was published in the September 2018 issue of SeaPower Magazine, in the Historical Perspective section, under the title "Weapon of Choice: The Navy's Role in Trinidad's Battle for Independence". (www.seapowermagazine.org)

[2] Wikipedia: Demographics of Trinidad and Tobago **https://en.m.wikipedia.org/wiki/Demographics_of_Trinidad_and_Tobago** (Consulted August 10, 2022)

Baptists) as well as Roman Catholics. Hindus account for 20.4 percent, Muslims for 5.6 percent. There is an Afro-Caribbean syncretic faith, the Orisha faith (formerly called Shango) with 1 percent and there are Rastafari with 0.3 percent. The "Other Religions" category accounts for 7.0 percent and "None/not shared" for 2.5."[3] Hence, these significant differences, in race and religion, offered no unifying structure to achieve a national cultural identity that could enable their pursuit for independence.

The U.S. military hegemony experienced by Trinidad was not exclusive to them, but rather it was a common practice that was replicated throughout the Caribbean islands, presented here within the context of WWII. Some islands have endured this state of affairs since the Spanish American War in 1898, as was the case of Puerto Rico. In Trinidad's case, it was an invariable regime of double consequence, given that they had to endure concurrent colonialism under both British

---

[3] Wikipedia: Demographics of Trinidad and Tobago **https://en.m.wikipedia.org/wiki/Religion_in_Trinidad_and_Tobago** (Consulted August 10, 2022)

governance and the gentrification by American military rule. Undoubtedly, an apparent grim situation to overcome under any circumstances, particularly in a country with diverse cultural and religious groups.

Its strategic position, located at the southernmost entrance of the Caribbean Sea, and in almost direct line to the Panama Canal, bestowed upon it an important large scale military role for both U.S. Navy and Army Air Force operations.

With World War II taking its toll on the allies in Europe, and German U-boats encroaching upon the Caribbean Sea, the English, French and Dutch governments petitioned the United States to protect its Caribbean possessions. Thus, in the wee hours of the morning of October 10, 1940, Trinidad's population awoke to see an imposing US Navy ship setting anchor in their Chaguaramas harbor. Casting a cloud over what would have soon been the joy of autonomy from the English Empire. The ship was the USS Saint Louis, a U.S. Navy, Brooklyn-class light cruiser full of sailors. They would soon learn that their dreams of autonomy were to be short lived. This, of course, ruffled a few feathers. Much to the colonial governor's chagrin, the U.S. armed forces were not going to be under his command. The island elite voiced their dissent, and the academics and nationalists resented the occupation because of what it represented to their anti-colonialist struggle. The population in general took a wait-and-see posture. The Navy impacted the population in many ways: their customs, culture, and economy. The U.S. sailors defied British rule with impunity, which emboldened the local men to do likewise. Sailors visited the bars, fraternized with the women, and socialized with the male population in the pubs.

While a sizeable portion of the middle class (i.e., accountants, teachers, government workers) were busy enjoying the newfound riches with the "*Yankee*" dollar during the construction of U.S. bases, the battle of Trinidad's independence fell upon the shoulders of the lower class and the less fortunate. It was to this group of destitute and marginalized citizens to whom the nationalists and the academics had to appeal to in order to mobilize an independence movement. This undertaking of mobilizing these citizens, befell primarily upon Dr. Eric Williams, who later became Trinidad's first President. As well as, Carlisle Chang, a talented musician, and Sylvia Chen, a ballerina by profession, and a fierce advocate for independence. Each used a variety of mediums at their reach to motivate the relegated population and whatever was left of the middle class, to join the independence crusade.

World War II made its presence felt in the Caribbean on the morning of February 16, 1942, when a German U-boat attacked Aruba's oil refinery. This did away with any opposition of the U.S. forces' presence in the Caribbean. Its sister island, Curaçao, was the largest high-octane refinery in the world, so it was of significant importance to protect these facilities. Of equal consequence were the oil deposits of Venezuela and Trinidad.

Population Displacement

Until the beginning of the twentieth century, Europe had been the cradle and icon of all things political, economic, and cultural. Three events set the stage for a new hegemonic world order by the U.S. These were: the Spanish-American War (April 21, 1898 – December 10, 1898), the First World War (July 28, 1914 – November 11, 1918), and lastly the Second World War. This allowed the United States to influence and dominate many countries around the world politically, militarily, and economically.

The U.S. spared no expense, and no effort was too little, to accomplish this goal. Over eight-hundred (800) U.S. military bases were built during WWII. Several existing bases, such as the one in communist Cuba, in Guantanamo Bay and the island of Puerto Rico, were expanded. These events had an enormous impact in geopolitics, which gave the United States great influence over many countries, at a time when the British Empire was beginning to cede their colonies. During the decade of the 1960s, many countries gained their independence from

Great Britain, which resulted in an increase in United Nations membership from fifty-one (51) countries to the current one-hundred and ninety-three (193) member nations.

Since its independence, United States' political, military and economic hegemony had always been of global design, far beyond its thirteen colonies. In his inaugural address to the nation, President George Washington referred to the new nation as a nascent empire, thus adopting a vision of expansion beyond the boundaries of the thirteen colonies. "*The farsighted Washington knew this as well, calling the United States at various times a "nascent empire", "rising empire", and "infant empire". To be sure, an empire can be at odds with liberty. The point here is that Washington had a sense that America was destined to be a great power—and indeed a global power.*"[4]

Under this premise, the Monroe Doctrine (1823) and the Manifest Destiny (1840) were written with the

---

4 Aha! Beyond News. **https://aha.world/president/george-washington** (Consulted July 7, 2015)

design to justify its expansion, political influence, and economic control of the western hemisphere. These two documents, without a doubt, expressed its expansionist objectives and divine right to do so. The Monroe Doctrine, written by James Monroe and John Quincy Adams (later to become the 6th President of the United States), was an address to congress meant as a warning to European powers that the western hemisphere was the United States' sphere of influence, and that they ought to steer clear from it.

Whereas, the Manifest Destiny has its roots since 1630s, when Puritan John Winthrop put forward a sermon titled "*City Upon a Hill*" depicting that the colonies had been chosen to spread democracy throughout the new world. Almost 200 years later, this theory, or philosophy, was re-inspired and written by John Louis O'Sullivan, a newspaper editor and columnist. He and his followers spread the belief that the United States had the God given "*destiny*" to conquer lands towards the western United States and spread their social and political beliefs. It was their view that they had the moral mission to do so, that they were virtuous, and that God had given them this undertaking to spread democracy throughout the continent.

The Monroe Doctrine establishes quite clearly the ambit and influence the US had over its neighbors and used it to put other countries on notice by dictating and limiting any action or interaction with any country within the western hemisphere by a foreign power. This came about at the time when Spain's influence and control in the new world was beginning to wane and ceased to present a threat of any import with which the United States had to deal with in any significant way.

On a broad scale, this doctrine establishes that any intent of colonizing or occupying any portion of the western hemisphere, other than those already in existence, by any other nation, shall be deemed as a direct aggression against the United States. It further states, that existing possessions by other nations are not transferable to non-other than the United States. This legitimized continued expansion by the U.S. The Monroe Doctrine has endured several amendments since its inception, and it has been executed by several presidents. In 1962, when the USSR attempted to establish military bases and install nuclear weapons on the island of Cuba, President John F. Kennedy successfully ordered a naval blockade, preventing Russian ships from reaching the island. Whereas, the Manifest Destiny, which started circulating in 1840, articulated that the US was *"predestined"* to expand its democratic and capitalist philosophy throughout the Americas. It also embraced the addition of the Oregon and California territories, as well as, Cuba and portions of Mexico as part of the union.

Soon after the swift removal of the Spaniards from the Pacific theatre and the Western Hemisphere during the Spanish-American War, and with the opening of the Panama Canal (August 15, 1914), United States launched a steadfast political and commercial presence throughout Latin America and the Caribbean. That same year, World War I broke out, which resulted in the destruction of Europe, thus giving the United States the opportunity to significantly increase its political and commercial influence in Europe. With the outbreak of World War II, in 1939, the U.S. military put into motion its world power ambitions and started building military bases throughout the western hemisphere, the Caribbean, and the Pacific.

Initially, there was no hurry in completing the construction of these bases. The attack on Pearl Harbor (December 7, 1941), though, changed all that and the afterburners were lit on all construction projects. During the Second World War, four hundred (400) bases were built on the islands of the Pacific and Atlantic oceans, and the Caribbean Sea, with the intention to form a defensive umbrella around the Panama Canal. With the attack on Pearl Harbor, this canal suddenly became the most protected and important real estate in the world. An attack on either side of the canal by Japan's I-boats in the Pacific, or by Germany's U-boats in the Caribbean, would have devastating results of epic proportions to both the military and commercial shipping traffic that flowed through it.

Between the periods of 1898 and 1939, the United States set into motion its geopolitical, commercial, and military objectives for the Western Hemisphere and the European continent. Once it decided that it wanted total control of the Western Hemisphere, the U.S. declared war on Spain in 1898, in what became known as the Spanish-American War - victory was swift. It took possession of Cuba, Puerto Rico, Guam, and the Philippine archipelago.

Eventually, Cuba and the Philippines gained their independence, whereas Puerto Rico and Guam became U.S. Territories.

With the opening of the Panama Canal, on August 15, 1914, the United States increased its military and commercial presence on both sides of the canal and invested heavily in the construction of military bases in the Caribbean and the Pacific. Hegemony brings about vast changes to any country or community, especially when it is pursued by a military occupation. Trinidad was no exception to this. Trinidad was soon the focus of Naval and Army Air Force support operations of significant scale.

All air and marine traffic with origin from Central and South America headed towards the United States or Europe were to first land or anchor in Trinidad for inspection.[5] Therefore, construction of military bases in the Caribbean had the singular objective of consolidating

---

[5] UKEssays. Trinidad and Tobago by World War II History Essay. Published 23 March 2015 **https://www.ukessays.com/essays/history/trinidad-and-tobago-by-world-war-ii-history-essay.php** (Consulted February 1, 2017)

the control of entry, thoroughfare, and exit of air and marine traffic throughout the Caribbean basin.[6]

The U.S. Navy had total control of all corridors, guaranteeing commercial and political influence throughout the region. The Panama Canal offered access control to the Pacific and Atlantic oceans. The gulf shores of Central America and the Florida Straits offered command of the Gulf of Mexico. While the Yucatan Canal, Puerto Rico, and the Lesser Antilles offered dominion of all cabotage in the Caribbean Sea.

U.S. Military hegemony of the Caribbean was with total disregard to land ownership, established system of government, or customs and culture. Land expropriations were the order of the day, evictions were demeaning and degrading. The modus-operandi were the same island after island. Hegemony and gentrification were not subjects for discussion, nor concern.

---

[6] The Caribbean basin consisted of (in alphabetical order): Bahamas, Barbados, Belice, Bermuda, Colombia, Costa Rica, Cuba, Dominica, El Salvador, Granada, Guatemala, Haiti, Honduras, Jamaica, Mexico, Panama, Puerto Rico, Dominican Republic, Saint Kitts and Nevis, Saint Vincent and the Grenadines, Saint Lucia, Trinidad and Tobago, and Venezuela.

Initially, opposition and resistance of the armed forces occupation emanated from the ruling class. Government entities sought to continue its existing form of government and had no desire to cede any of its powers to the U.S. military. The existing colonial hierarchy offered the minority white population the means to continue to subdue the overwhelming black population. The arrival of the U.S. military meant that the colonial governor will no longer have unlimited control of its justice system, commerce, customs, military powers, and most of all, control of the population and their actions. Academics and nationalists also made their opposition to the U.S. military occupation known because it now meant having to tolerate a colonial and patriarchal government, as well as an unknown military form of administration.

Particularly having had no advance notice, no period of transition, nor plebiscite, the people of Trinidad now had to deal with two concurrent colonialist governments. The Trinidad bourgeois, on the other hand, did not hesitate to display signs of resistance once they realized in what ways their lifestyles would be affected. The U.S. military's high demand of labor for the construction of bases and infrastructure, as well as operations of these installations, would result in a massive disarticulation of the local workforce that the government and bourgeois depended on (i.e., domestic workers, office workers, accountants, teachers).

U.S. Commercial interests with Trinidad dates to the 1870s when the U.S. commenced buying their cane sugar. With World War I, and with the advent of beet sugar in Europe, the British Empire ceased to import cane sugar from its colonies in the Caribbean. Cane sugar could not be shipped to Europe due to the U-Boats sinking of ships in the Atlantic. Moreover, beet sugar was much easier to produce and process, and yielded greater amounts of sugar per acre. Hence, the return on investment of beet sugar was far more superior. Seeking to establish a much larger and significant foothold in the region, the U.S. steps in to purchase almost the entirety of its cane sugar produced in the Caribbean basin. Thus, establishing a very close business relationship with the islands. With the advent of World War II, and the arrival of U.S. forces to Trinidad, one of their main hires for construction projects were sugarcane farmers, consequently crippling the islands' sugarcane industry.

When the U.S. Navy and Army arrived in 1940, the island of Trinidad had already been seeking sovereignty from the British Empire for several decades. The island's intellectuals, nationalists, and academics, would publish essays and editorials, and give impassioned anti-colonial speeches emphasizing the importance to unite as a society in order to better arrest the prolongation of their current colonial status – all to no avail. It was a difficult undertaking to accomplish considering Trinidad's varied cultures and religions, each pulling towards their own objectives and concerns. All the while, forced to adhere to the established social structure and behavior imposed by the British colonizers. Although the population had begun to view their cultural identity vastly different, they were still experiencing difficulties in identifying and giving shape to their own national cultural identity that would provide them with a common purpose as a society.

The arrival of the U.S. Navy to the shores of Chaguaramas, provoked an increased anti-colonial sentiment the intellectuals were eagerly seeking for its people to embrace. The U.S. occupation of Trinidad impacted the colonial government by essentially disregarding, or ignoring its customs and the established norms it had imposed on the population. This defiance towards the colonial government coming from U.S. sailors added an element of admiration on the Trinbagonian population, by realizing that they too can challenge the established colonial British rules with impunity. The U.S. military affected Trinidad's society, its customs, its culture, and its economy in many ways.

To better understand the actions, interests, and intentions the United States had for the Gulf of Mexico, the Caribbean, and the Western Hemisphere as a whole, it is necessary to delve into the ideologies of the U.S. government, the press, and the American population of 1898. They were eager to strip Spain from its Caribbean, Pacific, and Latin-American possessions, and rid Spain from its backyard to begin establishing a new world order. At a time when other empires were doing away with their colonial possessions, the U.S. was eager to expand its influence and acquire land throughout the world, both militarily and commercially.

In 1897, U.S. Senator Albert Beveridge said "*American factories are making more than the American people can use; American soil is producing more than they can consume. Fate has written our policy for us; the trade of the world must and shall be ours.*" [7]

Adding to the fray, The Washington Post published an editorial, in 1898, invoking the national mood and its imperial aspirations, "*A new consciousness seems to have come upon us - the consciousness of strength - and with it a new appetite, the yearning to show our strength ... Ambition, interest, land hunger, pride, the mere joy of fighting, whatever it may be, we are animated by a new sensation. We are face to face with a strange destiny. The taste of Empire is in the mouth of the people even as the taste of blood in the jungle.*" [8]

---

[7] Howard Zinn. "History is a Weapon: A People's History of the United States." **http://www.historyisaweapon.com/defcon1/zinnempire12.html** (Consulted November 21, 2012)

[8] Howard Zinn. "History is a Weapon: A People's History of the United States."

http://www.historyisaweapon.com/defcon1/zinnempire12.html (Consulted November 21, 2012)

The Spanish-American war (April 21, 1898 - December 10, 1898), was instigated by the sinking of the USS Maine in Havana's harbor, in Cuba. It was anchored there to assist in the evacuation of U.S. citizens and for national security purposes. With "Remember the Maine! To hell with Spain!" was the motto the press would use to garner support for the war.[9] Having won the Spanish-American War in just under four-months, the United States vastly increased its captive market for its goods and services throughout the western hemisphere virtually overnight.

World War I (1914-1918) can be viewed as the second phase of U.S. world dominance and influence. Although President Woodrow Wilson had assumed a neutral posture avoiding entering the war, the U.S. eventually had to declare war on Germany due to the heavy losses in sunken ships by the U-Boats. Meanwhile, he focused on expanding its commercial hegemony, sow its economic policies, capitalism, and democracy throughout

---

[9] Frankie Witzenburg. US Naval Institute: 'Remember the Maine, to Hell with Spain' The role the Maine played in the build up to the Spanish-American War. **https://www.usni.org/remember-maine-hell-spain** (Consulted April 21, 2013)

the European continent. World War I impacted Europe's political structure, by dismantling its economy, and obliterating its infrastructure. This allowed the United States to offer its loans, products, and services the European continent so badly needed for its reconstruction. This came with along with U.S. political and military influence. Then, with WWII, the U.S. consolidated its influence over the western hemisphere and the European continent, by expediting its military presence throughout the regions.

Due to its geographical position, less than seven-miles for the northeast coast of Venezuela, and once forming part of the South American continental shelf, Trinidad enjoys of abundant natural resources that were essential for expanding the infrastructure projects the United States was undertaking. Trinidad has bitumen, petroleum, natural gas, and quarries, a vital source of raw materials needed by the U.S. military for the construction of the bases on the island. In the 1880s, businessman Amzi Lorenzo Barber realized the great potential in Trinidad's bitumen for the construction of roads in the United States and was able to secure exclusive rights to the asphalt deposits. Mr. Barber was dubbed as "*The Asphalt King*".[10]

The U.S. started building its sphere of influence in the Caribbean by using its military as its hegemonic agent. They neutralized the colonial government oppositions and grievances by simply disregarding their leaders and their demands. Europe was facing a perilous, life or death, situation with the war being fought on its own soil. They were desperately fighting against a well-trained, well equipped, experienced army, able to fight concurrently in several fronts, terrains, and weather conditions. Hence, the European metropoles had neither the resources, nor the capabilities to tend to their possessions in the far reaches of the Caribbean. As a result, the English, Dutch and French governments requested the U.S. government to protect their colonies.

As part of the agreement with Prime Minister Winston Churchill, the United States government entered

---

[10] The House History Man: The "King of Asphalt" – Washington's Amzi Barber, **Published February 13, 2012,** https://househistoryman.blogspot.com/2012/02/king-of-asphalt-washingtons-amzi-barber.html (Consulted **September 21, 2014)**

into a ninety-nine-year (99) lease agreement of Britain's eight Caribbean possessions in exchange for fifty decrepit destroyers, known as the *Destroyer for Bases* deal. Churchill had been asking for destroyers since he took office during the Battle of Dunkirk, which perfectly aligned into the U.S. plans of world dominance. With Great Britain's diminishing world influence, and no longer masters of the high seas, the United States was now placing itself into the forefront of the new world order. As in WWI, the U.S.' position towards WWII was one of neutrality. Even so, the U.S. military had had some skirmishes with both the German and Japanese military prior to the attack on Pearl Harbor.

In 1937, the United States commenced construction of military bases on both the Atlantic and Pacific oceans without any haste. When Japan attacked Pearl Harbor, Japanese Admiral Isoroku Yamamoto said *"I fear all we have done is to awaken a sleeping giant and fill him with a terrible resolve."*[11] As a result of the strike, the United States shifted gears prompting military base construction at a fevered pitch throughout the globe. The total expenditure in Trinidad alone was forty-five million dollars ($45,000,000) for the construction of three bases and surrounding infrastructure, including roads to the airport, and an access road to Maracas Beach. The three-bases that were constructed in Trinidad were: U.S. Naval Base Chaguaramas, Waller Army Air Force Base, and Carlsen Army Air Force Base.

Trinidad's colonial governor, Sir Hubert Winthrop Young, consistently refused to cooperate with the war effort, and vigorously declined to pay for the construction and maintenance of these roads. The slightest resistance on behalf of any colonial governor that would impede the progress of the war effort would be immediately reported to the English government. The following day the Colonial Governor would hastily be in sync with the program or

---

[11] Awakening the Sleeping Giant: The Birth of the Greatest Generation, Published December 6, 2013, **https://blogs.va.gov/VAntage/11713/awakening-the-sleeping-giant-the-birth-of-the-greatest-generation/** (Consulted November 30, 2015)

shipped back to England. Some found themselves escorted to a ship headed back to England to never return.[12] The total expenditure for the construction of the Atlantic and Pacific bases was two-billion U.S. dollars.[13]

The colonial government of Trinidad felt neglected when it was given a mere four-day notice of the pending arrival of U.S. forces. This resulted in colonial government resistance and increased local nationalism on behalf of the elite. Other sectors of the population, either did not express much objection or assumed a wait-and-see attitude. Once the general population realized that the arrival of the U.S. Forces was going to create good-paying jobs, any opposition to their presence quickly faded. The colonial government was in a state of consternation because they were now playing second fiddle to a junior U.S. military officer.

---

[12] Neptune, Harvey R., *Caliban and the Yankees*: *Trinidad and the United States Occupation* Chapel Hill: The University of North Carolina Press, 2007, 92

[13] The Navy Department Library. Building the Navy's Bases in World War II, Vol. I p.9

The U.S. military injected a massive amount of capital and manpower to retrofit the Caribbean islands for the purpose of fighting the war against the axis forces. U.S. military bases were built throughout the island of Puerto Rico, including its sister islands of Vieques and Culebra. It also included bases in Cuba, Trinidad, Guyana, Britanica, Jamaica, Bermudas, Nicaragua, Panama, Surinam, Aruba and Curaçao, Honduras, Colombia, Uruguay, and Brazil. Of particular importance, Trinidad and Panama played a central role in controlling the access, and the inspection of, all aircraft and ships headed towards the United States.

The U.S. military gentrification of Trinidad is a story that resembles that of most Caribbean islands during WWII. There were massive amounts of dispossessing land owners of their properties, often with little or no remuneration. On many occasions, this occurred with the consent of the metropole and the complicity of the colonial government. Half-truths were announced and the local police was used to enforce the removal of property owners and attenuate any uprising of the population. In some cases, like that of Trinidad, even the colonial government had issues in the way things were done to accommodate the demands of the U.S. military but had little power or say on the matter.

Land expropriations impacted the deprived and most vulnerable segment of the population of their only means of livelihood, for it was land used for agriculture. Farming was mostly sugarcane, which was a much sought-after commodity in those days and represented a considerable amount of the island's Gross Domestic Product (GDP). These farmers had to seek a new form of earning their wages. In other instances, land owned by the elite, where they had built their weekend cottages, were taken away. Trinidad fishermen were ousted from their ports and were denied access to vast portions of their favorite and most productive fishing spots. They now had to navigate greater distances to fish.

Many of the farmers and fisherman found work in the expedited U.S. military base construction efforts. A total of ten-thousand (10,000) Trinidadians were hired. By paying at U.S. standard rate, local laborers became scarce and expensive, and the remaining workforce started demanding better job conditions from their bourgeois bosses. The bourgeois expressed their discontent to the colonial government, who in turn, at the suggestion of the elite, considered establishing a compulsory labor conscription system, as the one put into effect in Puerto Rico. It was never put into practice in Trinidad. "*Trinidad's agricultural interests cried that the colony crawled with lazy, immoral, and even criminal, young working-class males. "Hooliganism", their chorus went, was holding the society hostage, and conscription was the only effective policy response.*"[14]

The military's apathy towards the colonial government, their demands, and their English ways, emboldened the local population to shoulder the same position, to voice their concerns, and test the regime's resolve. Hence, it awoke in them a defiance they didn't realize they possessed and never ventured to manifest – the right to express their demands on equal footing, and have a say on their destinies.

---

[14] Neptune, Harvey R., C*aliban and the Yankees*: *Trinidad and the United States Occupation* Chapel Hill: The University of North Carolina Press, 2007, 92.

To further their hegemonistic position, the U.S. military chose to completely ignore the local salary scale, and equaled the workers' pay to that of their U.S. counterparts. This in exchange provoked the local workforce to demand better pay and improved workplace conditions from the government and the bourgeois bosses. A rare joint outcry of both the colonial government and the population, against the military, took place when the Navy blocked access to one of Trinidad's gems, Maracas Beach in the northern shores of Trinidad. The Navy caved in to their demands, and constructed an access road to the beach.

Revelers dressed as sailors during carnival in 1946

Sailor Mas - Traditional Mas Archive

Consulted October 10, 2022

US Navy sailors in Puerto Rico

Trinidad and Tobago During WWII. - ww2 post - Imgur

Consulted October 10, 2022

US Naval Officer Lt. Dillman with Trini girl

Trinidad and Tobago Archives - CARIBBEAN.MEMORY.PROJECT (caribbeanmemoryproject.com)

Consulted October 12, 2022

1940s Panyard – practicing for Carnival under the shade of a breadfruit tree

Trinidad and Tobago Archives - CARIBBEAN.MEMORY.PROJECT (caribbeanmemoryproject.com)

Consulted October 12, 2022

Military Road in the Northwestern peninsula of Trinidad (Chaguaramas)

Constructed by the US Navy Construction Battalion (Seabees) during the US occupation of the island in World War II

Seabees Museum (Consulted October 12, 2022)

US Navy Seabees constructing the access road to Maracas Beach for the local population and tourists

Seabees Museum (Consulted October 12, 2022)

US Navy base construction project in Chaguaramas, Trinidad

Courtesy National Archives, photo no. 71-CB-141P-15

US Navy base construction project in Chaguaramas, Trinidad

Courtesy National Archives, photo no. 71-CB-142T-9

US Navy base construction project in Chaguaramas, Trinidad

Courtesy National Archives, photo no. 71-CB-141F-2

Western Atlantic and Canal Zone Defense Area

Building the Navy's Bases in World War II, Vol. II, Chapter XVIII
Bases in South America and the Caribbean Area, Including Bermuda, Part I -- The Caribbean Area, Page 2

HyperWar: Building the Navy's Bases in World War II [Chapter 18] (ibiblio.org)

South American wildlife in Trinidad

See Yourself - Angelo Bissessarsingh's Virtual Museum of T&T - TTT News

Consulted October 12, 2022

South American wildlife in Trinidad

www.britannica.com/place/trinidad-and-tobago

Consulted October 20, 2022

Facebook page: Wildlife and Environmental Protection of Trinidad and Tobago - WEPTT
Consulted: November 23, 2022

Similar fuel and oil drums were taken to Trinidad by the US Army and US Navy

Stock Photos, Vectors and 3D Illustrations - Unlimited Downloads - 123RF PLUS

Similar oil drums were taken to Trinidad by the US Army and US Navy

WWII Crates, Boxes and Containers (questmasters.us) (Consulted October 12, 2022)

La Brea Pitch Lake, Brighton, Trinidad – 1988

Photo taken by the author

La Brea Pitch Lake 1988 – Tour Guide taking pitch seeping through the ground on a stick

Photo taken by the author

Holiday Inn Hotel, downtown Port of Spain – 1988

Photo taken by the author

Central Bank of Trinidad and Tobago – 1988

Downtown Port-of-Spain

Photo taken by the author

View from hotel window of downtown Port of Spain,
Trinidad and Tobago – 1988

Photo taken by the author

Trinidad Oil Platform

www.britannica.com/place/trinidad-and-tobago

Consulted October 20, 2022

On February 25, 2025, the government of Trinidad and Tobago officially decided to change their country seal. In exchange for the three (3) ships in which Christopher Columbus came to the new world, it now displays the steelpan – their national instrument.

Destroyers for Bases

Sir Winston Churchill was famous for his fiery speeches meant to rise the moral of the members of Parliament and, more importantly, that of the British people and subjects. The *"We shall fight on the beaches"* address to the House of Commons of the Parliament of the United Kingdom, given after the Dunkirk evacuation of 4 June 1940, was one such speech. It was also meant for Adolf Hitler's ears.

German ground forces and panzer corps surrounded the British Expeditionary Force (BEF), Belgian forces, and French armies on the beaches of Dunkirk. The only means of escape was by boat. British naval forces sent every destroyer they had at their disposal to evacuate the allied forces from the beach, but these were quickly put out of commission by Hitler's Air Force, the Luftwaffe. Churchill was in a very difficult position and quickly running out of options. He could not expend with the loss of more men and could not afford to lose the few destroyers that remained. Operation Dynamo (May 26 –

June 4, 1940), what was dubbed as the "Little Ships of Dunkirk", consisting of a private armada of over eight-hundred (800) British merchant ships, fishing boats, pleasure crafts, yachts, and lifeboats, saved the lives of three hundred and fifty thousand (350,000) allied men. All the same, tens of thousands of lives were lost.

Although President Franklin D. Roosevelt leaned towards helping Sir Winston Churchill in his war efforts, the U.S. Congress and the American people were reluctant in having anything to do with the war. To make matters worse, the manufacturing industry held a grudge against President Roosevelt for having imposed, what they considered, an onerous tax on them and resisted any request from President Roosevelt to convert their factories into a war effort machinery. Great Britain was short on cash, and its gold reserves were dwindling. Legally President Roosevelt could not sell weapons of war to any European country, but the law had a loophole – it mentioned nothing about lending or leasing.

The U.S. had a pressing need to expand its military presence in the Caribbean, while at the same time Great Britain had an urgent need to increase its naval fleet, each to counter the German offensive on either side of the Atlantic. *The Lend-Lease Act* of March 1941, also known as *"An Act to Promote the Defense of the United States"* allowed the President to enter into an agreement with Great Britain. Thus, circumventing Congress' opposition of selling instruments of war to the British. With this agreement, the U.S. would lend Great Britain 50 decrepit destroyers for a ninety-nine (99) year lease of its eight Caribbean islands. These islands were: Trinidad, Barbados, Bermuda, Bahamas, Jamaica, Antigua, Santa Lucia, and the region of British Guyana. The islands became known as *"The Destroyer Bases" and* were subjected to dual subservience by two patronizing world powers.[15]

Of the eight British islands, Trinidad's dispossession would prove to be the most consequential. "*Meanwhile, the German submarine campaign was being prosecuted with telling effect. Britain had been severely affected by the general attrition of operations at sea,*

---

[15] Lend-Lease Act, History.com Editors, Published November 4, 2019, **https://www.history.com/topics/world-war-ii/lend-lease-act-1** (Consulted February 10, 2022)

particularly in loss of destroyers. The United States needed extra bases to consolidate its defense in the Caribbean. Accordingly, both governments entered into negotiations which culminated in the "Destroyer for Bases" agreement, signed September 2, 1940. Britain received fifty over-age [sic] destroyers. The United States received the right, under a 99-year lease, to construct bases in eight British possessions (. . .)"[16]

Germany entered into an alliance with Italy (1930) and Japan (1935), forming the axis alliance. By 1939, United States was well aware of Germany's activities and degree of influence the Nazis were spreading in the Caribbean and Latin America. Juxtaposed with Hitler's victories in Europe, it increased U.S.' interest in establishing and expanding their military presence in the western hemisphere. As a result, the U.S. signed bilateral defense agreements with countries all along the Atlantic coast, from as far north as Terranova, to Argentina in the south. Thereby, establishing a defense shield against Nazi U-Boats all across the Atlantic seaboard. Should the Nazis

---

[16] The Navy Department Library. *Building the Navy's Bases in World War II,* Vol. I p. 4/171)

manage to establish a foothold in the Caribbean, it would place the U.S. in a precarious geopolitical position. It would also endanger the security of the 48 contiguous states, endanger its marine traffic to the U-Boat's crosshair, and put in jeopardy its influence in the western hemisphere. This was an extremely possible scenario with enormous consequences. Germany had already overpowered France and the Netherlands. Had it defeated England, it could very well have established permanent military presence in the Caribbean islands of these metropoles.

WWII re-introduced a paradigm few had accomplished throughout history, the ability to sustain battle in various fronts, at sea, across oceans, and in different continents. The Nazis demonstrated that they had the means to weather such conflicts very effectively. For said reason, the U.S. had the need to expand its support bases for its naval and aircraft fleets. By 1939, annex bases of little importance had already existed. These required retrofitting to meet their new mission. There was the naval base in Guantanamo, Cuba (GTMO), a radio tower in San

Juan, Puerto Rico, and trivial military operations in Panama and St Thomas, USVI.

Such expansion had commenced by the end of 1939 without any particular haste.[17] The attack on Pearl Harbor changed all that, and the gears swiftly shifted into urgency mode. The construction projects were a colossal undertaking, which required massive funds assignments and manpower. Base constructions commenced on March 1, 1941.[18] "*In 1940 the Navy had no properly equipped advance base other than Pearl Harbor. In the succeeding five years the Bureau of Yards and Docks built or supervised the building of more than 400 advance bases for the Navy in the Atlantic and the Pacific areas, at a cost of $2,135,427,881.*"[19]

With Japanese subs lurking the Pacific waters and German U-boats now roaming the Caribbean, the

---

[17] Abbreviated as GTMO in official Naval Communications, and pronounced as GITMO

[18] The Navy Department Library. Building the Navy's Bases in World War II, Vol. II p. 4/171

[19] The Navy Department Library. *Building the Navy's Bases in World War II,* Vol. I p. 9/137

protection of the Panama Canal was supreme. An attack on the canal would be devastating to both commercial and military traffic. Hence, President Franklin D. Roosevelt asked the Joint Chiefs of Staff to draft a plan to counter any threat to the region. When Secretary of State, Henry Stimson, together with the Joint Chiefs of Staff, presented President with their defense plans, he approved it by writing *"Army/Navy O.K. F.D.R."* on the map. By pure chance, the word Army was written over the Isthmus of Panama, while the word Navy fell onto the Caribbean Sea. The admirals took advantage of this to claim the Caribbean as theirs.

The main design behind establishing military bases on either side of the Panama Canal, and throughout the Caribbean, was to cast a wide defense shield around the Isthmus to protect and ensure the unimpeded navigation of ship traffic against any Japanese and German submarine attack. A total of four-hundred (400) bases were built between the Caribbean Sea, the Pacific, and Atlantic Oceans. These had the purpose of providing logistical and operational support to a fleet of over

15,000 aircrafts and vessels. Given the sheer volume of commercial and military ships that traversed through the Isthmus, the Canal became the most defended and prized realty in the world. A successful axis attack on either side of the Isthmus would result in an economic and military catastrophe. Hence, with Panama to its west and Venezuela on its southern shore, Trinidad was converted into a base of major importance. "*Trinidad was to be organized as a subsidiary operating base, with major air facilities capable of supporting a large portion of United States naval forces. It was also to include training and advance-base operations projected to the south and east. Secondary air bases were to be located at St. Lucia and British Guiana, and emergency advance bases were to be placed along the northeastern coast of Brazil, each to be linked logistically to Trinidad and geared to the defense plan of that area.*" [20]

The residents of Aruba were awoken by an explosion in their bay on the morning of February

[20] The Navy Department Library. Building the Navy's Bases in World War II, Vol. II p. 3

16, 1942. Thus, making the island the first casualty of major importance of WWII in the Caribbean basin. The U-Boat sank seven ships, two of which were anchored in Saint Nicholas harbor. In a bold move, the Nazi submarine flaunted the allies by surfacing and launching two torpedoes at Aruba's Lago oil refinery. The fire ignited by the blast lighted the night sky that could be seen from its sister island, Curaçao. This placed Curaçao on high alert that an attack was imminent, allowing them to prepare and repel any assault coming their way.

Curaçao possessed the largest oil refinery operations in the world. Producing high grade fuel required for the aircrafts and ships in Europe and the Pacific. A commodity much desired by the Nazis. *"The battle of the Caribbean, heralded by the attack at Aruba, took the form of a sustained and extremely damaging U-boat assault against shipping. Its first victims were five tankers -four of them British ships and the other a Venezuelan- that were torpedoed and sunk during the early morning hours of 16 February 1942. Insult was added by the fact that two of the ships were sent down while lying at anchor in San Nicolas harbor, Aruba, by a U-boat that entered the anchorage, sank the two tankers, then came boldly to the surface and lobbed a few shells at the Lago oil refinery."*[21]

The German Army invaded the Netherlands on May 10, 1940. Venezuela had declared neutrality during the war but was nonetheless entertaining Nazi propaganda. Hence, Netherland was reluctant to the thought of accepting that Venezuela participate in any defense efforts of their Caribbean possessions of Aruba and Curaçao. Given that Curaçao's refinery was one of the few, and largest, producer of 100 octane fuel, it was of utmost importance to resolve this matter as quickly as possible. In fact, the Navy had created a team that dealt solely to the construction of similar facilities. *"Had it not been for some dissatisfaction on the part of the Dutch over command arrangements in the Far East and their unwillingness to permit Venezuelan participation in the defense of Aruba*

---

[21] Conn, Stetson, Engelman Rose C., Fairchild, Byron. Center of Military History, United States Army, First Printing 1964. *Guarding the United States and its Outposts, The Caribbean in Wartime*, Chapter XVI, p.423

and Curaçao, and for some reluctance on the part of the United States Government to press the issue until after the Rio de Janeiro Conference of Foreign Ministers, faster progress might have been made."[22] This gave rise to U.S. diplomatic initiatives, which resulted in the delay of defense preparations by the allies. "*The dispatch of American troops to Aruba, Curaçao, and Surinam, and the entry of the United States into the war as an associate of Great Britain and the Netherlands, raised a problem of command relationships not only between the Army and the Navy, but also between the United States and its allies.*"[23]

[22] The Navy Department Library. Building the Navy's Bases in World War II, Vol. II p. 23

[23] The Navy Department Library. Building the Navy's Bases in World War II, Vol. II pp. 99/137-99138

Cultural Exchanges and Conflicts

Hubert Young, Trinidad's Colonial Governor, was vehemently opposed to any U.S. military presence on the island. It significantly curtailed his authority. His pretensions were to integrate the U.S. military forces to his army and have full command and control. The U.S. Consul, the Generals and Admirals, informed the governor, in no uncertain terms, that the U.S. forces were not going to be under his command, nor that of the British Empire. He fiercely opposed losing Chaguaramas Bay to make way for the naval base. This area enjoyed of pristine beaches, lush portions of land, and bays. Aside which, it was where the politicians and the bourgeois maintained their summer villas, residences, clubs, and country homes. The Governor wasted no time in expressing his discontent of the occupation to his superiors. And did so at every opportunity. "*I have considered the whole question in Executive Council and am advised that the acceptance of the present proposals would undoubtedly cause great dissatisfaction among those elements of the community who would be deprived of existing amenities and might well, in view of this and of the siting of the proposed locations, lead to serious difficulties during the period of the lease."*[24]

The lease of the British Islands, granted to the U.S. government by the Monarchy, gave the U.S. military unfettered access to its lands and waterways. Residents, land owners, and businesses in Chaguaramas were evicted with no regard to their livelihoods. To add injury to insult, they did so with colonial government agencies spearheading their removal. Governor Young made an effort to prevent this mass confiscation of property and relocation of residents by denoting that the colonial government did not have the budget to undertake the construction of required infrastructures, nor that of its

[24] Directory of Research and Expertise. School of Advanced Study. University of London. Atlantic Archive. *UK-US Relations in an Age of Global War 1939-1945.* Trinidad Governor Sir Hubert Young on his objections to US proposals for US bases on Trinidad, including proposed alternatives, October 22 1940 **http://archive.atlantic-archive.org/193/** (Consulted March 3, 2012)

maintenance. Hence, the United States shouldered the bill, as well as construction efforts.

Another point of controversy was that the lease contract did not specify which entity had legal jurisdiction over a soldier or sailor's criminal conduct. The Colonial Governor had suggested a dual prosecutorial venue. The U.S. Consul and the U.S. military brass voiced their opposition and it was agreed that the U.S. military would retain all legal authority on such matters over their men. By 1942, Governor Hubert Young had amassed sufficient opposition to any and all U.S. military efforts that he was removed from office and shipped back to London on the first available vessel. "*Governor Young clashed with the Americans several times, not only on the issue of the villagers having to leave Chaguaramas, but on the question of the bathing beaches being put out of bounds to holiday seekers and ordinary villagers. He did not like the idea of the Americans having Chaguaramas and wanted them to develop the Caroni Swamp instead and establish a base there He was overruled and eventually sent home to England."*[25]

So far as the contracted workers were concerned, the first problems encountered by the sailors and soldiers during the construction of the bases and its operations were cultural-linguistic in nature. Despite Trinidad being a British colony, its population spoke a mixture of English with a local vernacular, or jargon, as a product of its diversity in races, religions, cultures, and subcultures. Not to mention a marked accent that made it difficult for U.S. military personnel to fully comprehend what was being said.

The U.S. military was given access to prime land without any reflection on how it may have impacted those who lived there. The U.S. Navy occupied a total of seven thousand nine hundred and forty (7,940) acres. The town of Chaguaramas alone measured three thousand eight hundred (3,800) acres, consisting mostly of private farmland. The population raised an uproar when the Navy

---

[25] John Boswell. Chapter 2 – 1941 – 1998 The Americans Triniview.. **http://www.triniview.com/Carenage-Chaguaramas/Chaguaramas2.html** (Consulted June 1, 2013)

blocked access to the gem of their beaches, consisting of pristine calm blue waters, known as Maracas Bay Beach on the northern shores. This beach attracted tourists and locals alike. To appease the situation, an adjacent road providing access to the beach was constructed by the Navy's Construction Battalion (CBs) – better known as the SeaBees. Notwithstanding the jobs created for the local population, this did not placate the people, nor the colonial government. Relations were strained and it always remained as a point of contention.

During base constructions, hostile situations ensued among the civilian workers mostly due to gang rivalries, guild complaints, and clashes over job openings. At times, these quarrels would turn violent and result in dismissals. "*(. . .) and had difficulty in understanding the English of the continentals, who had equal difficulty understanding them. There was a definite caste distinction, not only among the different races but among the different employment classifications. They were temperamental among their own groups, which often resulted in serious fights (. . .)"*[26]

Trini academics and nationalists had, for decades, drafted and gave every fervent speech they could conceive in the hope that it would provoke the people of Trinidad into identifying themselves as one, as countrymen, as a nation with a common goal in antipodal consonance to British rule. During the 1930s, they had realized that the country's needs were diametrically distinct to that of the British. A need which Great Britain could not properly address nor satisfy.

Labor disputes took place throughout the West Indian Islands at the height of the Great Depression in the 1930s, which gave credence to the thought of independence from the British Empire. Although the labor unrest was mostly due to wealth inequalities, it also voiced other grievances, such as, housing, education, healthcare, and other disparities. With the exception of Trinidad, who had oil, the rest of the islands depended mostly of cane

---

[26] The Navy Department Library. Building the Navy's Bases in World War II, Vol. II p. 24/235

sugar exports for foreign currency which suffered of many market price fluctuations. Strikes subsided at the start of WWII.

To achieve independence, though, it was essential that the nationalists and academics convey a message that would awaken the people's passion, a message that would resonate in their souls. They needed to come up with a concept that would marshal what it meant to be a Trini. They were not achieving it. If the people of Trinidad could not identify themselves as Trinis, there would be no sovereign Trinidad. The country was too divided into diverse races, cultures, religions, and political ideologies. They had to unite, they had to devise a message, a concept, that would allow them to envision themselves as a country.

It was not an academic, nor a nationalist, nor a politician who discovered the key to their unity. But rather, the essence to its national affirmation, was discovered by a young Trini. It resided in the fifty-five (55) gallon steel drum barrels that the U.S. military brought with them full of high-grade lubricants for their aircrafts. As they were discarded, the population would convert these drums into ovens to bake, barbecues, garbage bins and burning pits.

Trinidad's fate took a turn one day when a young man pounded away at the dents on the Navy's discarded oil drums. His fine-tuned ears discerned different musical tones gracefully flowing from the drums. From these barrels, the steelpan was forged and the steelpan was born.

Carlisle Chang, then a teenager, and Sylvia Chen, both born in Trinidad, and of Chinese descent. They are credited in having had a major influence in the development of Trinidad's national cultural identity that united and defined the Trinis as one culture.

Sylvia Chen was born in 1909 and had lived in England, Russia, China, and the United States. She was a ballerina by profession and worked for the Bolshoi Ballet, and did choreography for Hollywood. She was a passionate anti-colonialist, most likely influenced by her father who fought against the Japanese's first occupation of China in 1894. After a 14-year absence, Chen returned to Trinidad and started producing dance dramas that would convey to her audience the colonial government's abuses of her people. Chen published an op-ed in the local press imploring Trinis to come up with a concept that would distinguish them as a unique culture ". . . *develop something we can call our own.*"[27] Even a major detractor of Sylvia Chen, Albert Gomes, had to admit that she was able to attain a renewed interest amongst her people to see themselves from "*another perspective.*"[28]

Carlisle Chang, on the other hand, was a talented musician who was able to articulate the four elements that produced an unlikely cultural mosaic that transformed a fragmented society into a country that danced and sang to the same beat. Chang blended the calypso lyrics, along with the percussion qualities of the steelpan, to convey a unique musical expression which to this day represents their autochthonous culture. Together with their local dance and infused with its carnival spirit, it morphed into their newfound cultural expression. The people immediately embraced it, irrespective of their race, religion, culture, or political affiliation.

---

[27] Neptune, Harvey R., C*aliban and the Yankees*: *Trinidad and the United States Occupation.* Chapel Hill: The University of North Carolina Press, 2007, 47

[28] Neptune, Harvey R. *Caliban and the Yankees: Trinidad and the United States Occupation. "Chen had started something and perhaps has stimulated us to a greater interest in ourselves.",* Chapel Hill: The University of North Carolina Press, 2007, 47

"There was no carnival in Trinidad and Tobago from the years 1942-1945 because of the Second World War. The early steelband was not even called steelbands. Those first "instruments" consisted of such things as biscuit tins, dustbins, sticks, bongos, pots, and hub caps. The bands would have been called dustbin bands. Back in those early days you probably had to chain your dustbins or you would find them gone. You had just made an unwilling contribution to some dustbin band. Lol."[29]

These rudimentary instruments were replaced by the steelpan, which resulted in a natural infusion with the calypso lyrics that were used by the Trinis as an informal news source to inform the people of imperial actions against the country. Today, it is still used as a colloquial venue to express the people's feelings and gripes.

Although there were calypsonians that understood that there were potential entertainment opportunities in tourist who were willing to pay to listen to this newfound

[29] **Dr. Kim Johnson,** Director of the Carnival Institute of Trinidad and Tobago. "Mr. Nestor Sullivan. History of Steelband in Eras." YouTube video, 1:19:26. **https://youtu.be/pIGOSmvyBgI** (Consulted December 27 2016)

music form, there too were calypsonians eager to rid the U.S. military presence from their land. One such calypsonian was composer and singer, the Mighty Sparrow, who wrote the following calypso celebrating the base closures, titled "Jean and Dinah"

> *Well, the girls in town feeling bad*
> *No more Yankees in Trinidad*
> *They going to close down the base for good (. . .)*
>
> *Don't make no row, the Yankees gone, Sparrow take over now*
> *Things bad is to hear them cry*
> *Not a sailor in town, the night clubs dry (. . .)* [30]

Syliva Chen's preferred calypso singer-composer was Neville Marcano (1916 – 1993), better known as Growling Tiger.[31] He was a fierce critic of the gentrification and presence of the U.S. military in Trinidad. The displacement of the Trini population from

---

[30] Neptune, Harvey R. *Caliban and the Yankees: Trinidad and the United States Occupation.* Chapel Hill: The University of North Carolina Press, 2007, p. 47

[31] Growling Tiger. YouTube video, 2:59. https://www.youtube.com/watch?v=4acONMykrb4 (Consulted April 2, 2013)

their lands was without any regard for their family nucleus or how they earned their wages. In one of his songs, he highlighted the detrimental influence the dollar had on Trini society. He wrote the following calypso titled *"Money is King"*.

> *If a man has money today*
> *People do not care if he has cacobe [yaws, a tropical disease]*
> *If a man has money today*
> *People do not care if he has cacobe*
> *He can commit murder and get off free*
> *Live in the governor's company*
> *But if you are poor, people tell you "shoo"*
> *And even dog is better than you*
>
> *A man with money walks into a store*
> *The boss will shake his hand at the door*
> *Call ten clerks to write down everything*
> *Suits, hats, whiskey, even diamond rings*
> *Take them to your home on a motorbike*
> *You can pay the bills whenever you like*
> *And not a soul will ask you a thing*
> *They know very well that money is king (. . .)*[32]

The calypso composers possessed a keen sense of the islander's everyday reality and understood their wandering state of mind, and would express them in their

---

[32] Neptune, Harvey R. *Caliban and the Yankees: Trinidad and the United States Occupation.* Chapel Hill: The University of North Carolina Press, 2007, p. 189

songs. One such example is the song "*Hosein*", which highlighted the festive culture of the island Muslims, or "*Carib Woman*" which called attention to the declining native population. "*The group championed an aesthetic that was "an expression of [their local] environment", and tellingly, its inaugural exhibition in January 1933 inspired one reviewer to remark that the event evinced the emergence of an "amorphous consciousness of the island."*[33]

As the steelpan began to acquire acceptance amongst the population, they realized that it was an ideal instrument for their calypso songs due to its percussion qualities. It was an unquestionable match, which took on a greater meaning when it was integrated with the local dance and carnival. The means to its national edification was inadvertently discovered in these oil drums.

Their war involvement in the Caribbean impacted the islands in a variety of ways, but much more so in Trinidad by giving way to the discovery of the steelpan by

---

[33] Neptune, Harvey R. C*aliban and the Yankees*: *Trinidad and the United States Occupation.* Chapel Hill: The University of North Carolina Press, 2007, p. 45

the creative mindset of the Trinis and thereby paving the way to the affirmation of self-rule. They developed a way to fine-tune these barrels, and together with their calypso songs, what came to fruition was a proud and common heritage. The percussion quality of the steelpan, together with the colorful language of the calypso songs were a perfect match. It resulted into an unquestionable accomplishment. Later, they included their folklore dance and added these to their carnival parades. The original percussion instruments were largely composed of sticks, pots, pans, and grates. Trinidad and Tobago morphed into what it is today, and since have proudly been known as "Trinis" or the term preferred by the politically correct "Trinbagonians".

As soldiers and sailors were transferred to other bases around the world, they became the de facto ambassadors of Trinidad's music and culture. They would tell stories of their adventures in Trinidad and would play recordings they had made of the music. Thus, giving Trinidad's voice a world stage audience. The music was never accepted by the British Colonial Government, and was repudiated by the bourgeois.

Calypso music, of West African origin, was accompanied by very rudimentary instruments and was always used to express dissent and gripes against other religious and cultural groups or gangs, as well as the government and upper class. But when the steelpan music was discovered, it offered gangs a more colorful stage, so to speak, for them to express their people's rifts. Each neighborhood had their own rhythm and unique steelpan beat, as a means of gang identification. Calypsonians understood that with the advent of this new musical instrument, the structure of calypso music had enriched their genre significantly.

"Not everyone was into what would become our National Instrument – and it wasn't just the authorities. In the early days it was mainly popular among the youth. It was certainly a young people's thing. The same way that their parent's might not like Elvis Presley they did not readily accept their children being involved in steelband. Remember the pans back then did not look like the sophisticated thing it is now. In fact, back then if it was discovered that you were a steelband man

you would become an outcast. Calypsonians too were seen as outcasts.

Both were not held in high esteem for the reason that many calypsonians and panmen were either underemployed on unemployable due to a lack of education at the time. Though there were a few Calypsonians like "Atilla the Hun" and "Lord Executor" who had been to college but chose the life of a calypsonian despite the lack of status it brought.

No parent wanted their daughter's associating with such a person (Panman or calypsonian) and worse if they found out that their sons were interested in it. Clearly there was a generational gap.

To get a truer picture of what the coming of the Americans meant to Trinidad and Tobago you would have to look back at what was happening economically and socially before they came. Before the coming of the American soldiers, whites (the British) were seen as almost godlike. The Americans were far cruder and down to earth nothing like the British. With that in mind it made it easier to think of shaking off colonialism. It is interesting to note as well that America too had shaken off the yoke of the British.

I don't know if you were aware of this but during the period of which you speak and while Pan was still in its infancy there were violent clashes between rival steelbands. Cutlasses, stones, bottles, knives and razors were some of the weapons of choice. There were deaths and injuries. The authorities were not too pleased. Two of the main reasons for this it would appear were TURF and WOMEN. Bands from one area did not like bands from other areas invading their turf. Similarly, they did not like panmen from one area coming in to their area and trying to fraternize with their women. Because of the violence steelband and steelband men got a bad reputation. In time though the warring would come to an end."[34]

"As much as Trinbagonians welcomed the benefits brought by the Americans – they liked watching American films and it influenced the Carnival in terms of ideas for portrayals, names of steelbands. They copied and

---

34 **Dr. Kim Johnson,** Director of the Carnival Institute of Trinidad and Tobago. "Carlton 'Zlgullee'Constantine: Steelband Wars." YouTube video, 6:31. ***https://www.youtube.com/watch?v=uS_a7FQ16lU*** (Consulted December 27 2016)

mimicked the bad behavior of drunken sailors ashore. Thus, was born a specific group of sailors portrayed in Carnival known as "The Bad behaved Sailor". This type liked to roll around in dirty drains and dirty up their "uniforms" and smacking ladies on their bottoms.

Certain dance movements also came out of their contact with the soldiers. On Carnival Day groups of revelers dressed in sailor costumes would be seen with their arms around each way going to the left while they group in front would move to the right and so on. This was called "Rocking the boat. It mimicked the movement off the sailors on the boats. King Sailors (differentiated by the crowns on their heads) are known for their fancy dance movement, which takes years of practice to master. Each King Sailor has his own unique set of steps and foot movements. One old King Sailor of many years Mr. Ralph Dyett in an interview with me said that the sailor dance had some influence on Michael Jackson's moon walk. Jackson having visited Trinidad. Having watched him dance I can see how Jackson could have been impressed with this style."[35] *"A sailor dance grew from locals observing the unsteady walk of the sailors as they came ashore after long periods at sea. Mas players soon copied and elaborated"*[36]

The steelpan enjoys two notable distinctions: it is the only musical instrument discovered during the twentieth century, and second, it was constructed out of industrial waste created by the U.S. Navy. Today, there are 11 steelpan varieties. Without a doubt, the steelpan played an invaluable role in Trinidad's battle for independence. Today, the steelpan is used in carnivals in all corners of the earth, including in the far reaches of China.[37]

---

[35] **Dr. Kim Johnson,** Director of the Carnival Institute of Trinidad and Tobago. "Mr Austin Wilson – Sailor Mas" YouTube video, 6:10. ***https://www.youtube.com/watch?v=QDxQqoUdrjs*** (Consulted December 27 2016)

[36] **Dr. Kim Johnson,** Director of the Carnival Institute of Trinidad and Tobago. "Traditional Carnival – Fancy Sailor teaches Dance." YouTube video, 2:37. *https://www.youtube.com/watch?v=jDrHAHopjU8* (Consulted December 27 2016)

[37] "Chinese Carnival Cheng Du". YouTube video, 4:19. **https://youtu.be/Sh0TZX8XtM8** (Consulted June 15, 2022)

The Great Eviction

By 1946, once WWII ended with Japan's defeat (September 2, 1945), most military personnel were no longer needed in Trinidad and were transferred to other bases around the world after a 6-year occupation of the island. However, despite the drastic reduction of Naval operations in Chaguaramas, the U.S. did not transfer the facilities to the colonial government. Instead, it was handed over to other federal agencies for their use. One such agency was the National Oceanic and Atmospheric Administration (NOAA), and a U.S. Coast Guard Long Range Aids to Navigation (LORAN) station, among other federal agencies.

From a cultural perspective, both Sylvia Chen and Carlisle Chang, were central to Trinidad's new found identity. But there, too, was a charismatic political leader who possessed an unwavering will and foresight to deal with his country's struggle for independence and eviction of the U.S. military from the island. This was Dr. Eric

Williams, a historian, and a man who did not mince words, nor did his knees buckle when it came to fight for his country. He was instrumental in piecing together the relentless effort of depriving Great Britain and the United States of any further presence and governance of his homeland and his people.

The English Islands agreed that a united front would surely give weight to their independence cause. Thus, with Great Britain's backing, the islands formed the West Indies Federation on January 3, 1958, with its capital in Port-of-Spain, Trinidad. They decided upon a similar design used by Canada, known as the Confederation of Canada. In 1960, under the Federation's banner, Dr Williams, with the citizens of Trinidad by his side, marched to the Naval base to reclaim their land and make Chaguaramas the Federation's capital. Despite the torrential rain, the march attracted a massive turnout. The theme song for the march was "*Uncle Sam, We Want Back*

We Land'.[38] The Navy agreed to the meeting, and the flags of all those present in the meeting were flying in front of the building. These were, the United States, Great Britain, the Federation of West Indies, and Trinidad and Tobago. A second meeting was agreed to for the following month of February 1961, where all parties would review the lease agreement in detail.[39] The Federation had decided not to bring up the issue of having the Navy cede the base for fear of missing out on the two-million dollars ($2,000,000) annual fee they were receiving from the United States.[40] Dr Eric Williams was opposed to this because it would give the Navy reason to remain on the island.

To make matters worse, shortly before the Federation would gain independence, Jamaica resigned

---

38 MAWBY, SPENCER. "'Uncle Sam, We Want Back We Land': Eric Williams and the Anglo-American Controversy over the Chaguaramas Base, 1957-1961." *Diplomatic History*, vol. 36, no. 1, 2012, pp. 119–45. *JSTOR*, http://www.jstor.org/stable/44376139. (Consulted 10 Sep. 2022)

39 Neptune, Harvey R., C*aliban and the Yankees*: *Trinidad and the United States Occupation* Chapel Hill: The University of North Carolina Press, 2007, p. 192

40 Neptune, Harvey R., C*aliban and the Yankees*: *Trinidad and the United States Occupation* Chapel Hill: The University of North Carolina Press, 2007, p. 192

from the union due to philosophical and political differences, leaving Trinidad with seventy five percent (75%) of the organization's financial obligations. As a result, Trinidad withdrew its membership and Great Britain dissolved the organization on May 31, 1962.[41]

"As for Eric Williams, having attended Oxford University, belonged to the educated elite. What he noticed about the steelpan was its ability to bring all different types of people together. Eventually even middleclass college educated youth were drawn to the steelpan and started their own bands. Calypso had its roots in slavery and together with pan I think they were not so much tools of rebellion but that the climate at the time and the social conditions led to a perfect storm for such rebellion. Though they (Americans) contributed to the economy and the bases provided much needed employment the people exploited the presence of the Americans but as noted in several calypsos their presence also had its negatives. There was the rise in prostitution and resentment among

---

[41] Neptune, Harvey R., C*aliban and the Yankees*: *Trinidad and the United States Occupation* Chapel Hill: The University of North Carolina Press, 2007, p. 0192

the local male population who could not compete with the soldiers and their Yankee Dollars. In the end they were glad to see the backs of the Yankees as seen in one of the Mighty Sparrow's popular hits "Jean and Dinah."[42]

Since Trinidad's occupation had been an agreement between the U.S. government and Great Britain, it was Dr. Williams' understanding that it was incumbent upon Great Britain to evict the U.S. from their island. It was not until after Trinidad's independence, on August 31, 1962, that Trinidad was able to compel the U.S. government to sit at the negotiating table to discuss their departure. An effort that lasted 15-years. The U.S. centered their case on the ninety-nine (99) year lease contract to evade leaving the island. Dr. Williams' negotiation team made it clear that they were now dealing with the nation of Trinidad and Tobago and that, in essence, the agreement with Great Britain was now null and void.

Unlike other islands, Trinidad displayed a resourceful ability of converting the challenges they faced

---

[42] **Dr. Kim Johnson,** Director of the Carnival Institute of Trinidad and Tobago.

into their favor. After the Chaguaramas meeting, Dr Williams decided to go through the lease agreement with a fine-toothed comb, and focused on the fine print. Having carefully read the contract, he discovered that the document lacked the Seal of Your Majesty of the Colony of Trinidad required by the Inland Revenue Office of Property Registration. Neither did it have the necessary permits and legal stamps for the other bases and lands it occupied. This put into question the legal validity of the lease agreement, as well as any obligations Trinidad may have towards the United States.

Lastly, it also stipulated that any land transfer by the Royalty to a third party, must have the approval of the Colonial Governor, and that such lease cannot exceed 30-years. Any extension to the lease must be in blocks no greater than an additional 30-years, and cannot exceed by more than 3 consecutive extensions. The lease contract did not comply with any of the legal requirements.[43]

---

[43] Oxford Journals. Arts & Humanities. Diplomatic History, Volume 36, Issue 1, Pp, 119-145. **http://dh.oxfordjournals.org/content/36/1/119.abstract** (Consulted February 1, 2020)

Thanks to its steelpan music, Trinidad now enjoyed a worldwide audience and fame, and the United States could not just sweep this issue under the rug. With this information of breach of contract, together with the extremely low rent fee, Dr. Williams now had the attention of the United States government. The U.S. had no choice but to negotiate a new contract, cease and desist from any further Naval and Army operations, and return all seized lands. Nonetheless, the U.S. government retained the right to return in the event of a national security until the year 2040.

Dr. Williams knew all too well of his people's toil and sacrifices, and understood that it was they who had to direct his political platform. As a keen observer, he realized that calypso lyrics and steelpan music was pivotal in uniting his people behind the common cause of removing the U.S. military from their land. Having gained their independence from Great Britain, Trinis were living a new reality, they were embracing a new form of expression with great pride. He welcomed this musical paradigm into his political crusade and offered calypsonians a stage in which to express the people's views during Trinidad and Tobago's Independence Day festivities. Throughout the fête, Calypsonians would compete with songs that highlighted their unification regardless of race, religion, or culture.

Dr. Williams realized that they were now Trinis, or Trinbagonians, and that he now had to wage war against the U.S. Navy with their pan music and piercing calypso lyrics. Now, the voice of dissent toward U.S. occupation came not from the colonial government, the academics, or nationalists, instead it emanated from the locals and their steelpan music - their weapon of choice.

About the Author

Louis Miranda, an author, and historian, lives in Guaynabo, Puerto Rico. He joined the U.S. Coast Guard, in 1973 and was deployed to U.S. Air Force and U.S. Navy bases in Puerto Rico. Having traveled to Trinidad and Tobago for a span of over 30-years, and having served in the military, it was just natural to choose Weapon of Choice as the thesis for his master's degree.

This book is the author's contribution to Trinidad and Tobago's history and culture.

woc.thesis@gmail.com

ISBN 979-8-218-12548-6
90000>
9 798218 125486

www.ingramcontent.com/pod-product-compliance
Lightning Source LLC
LaVergne TN
LVHW020511100826
845148LV00003B/761